FARMING THE LAND

FARMING THE LAND

MODERN FARMERS AND THEIR MACHINES

by Jerry Bushey

Carolrhoda Books, Inc./Minneapolis

This book is available in two editions:
Library binding by Carolrhoda Books, Inc.
Soft cover by First Avenue Editions
241 First Avenue North
Minneapolis, Minnesota 55401

LIBRARY OF CONGRESS CATALOGING-IN-PUBLICATION DATA

Bushey, Jerry.
   Farming the land.

   (Carolrhoda photo book)
   Summary: Discusses modern farming methods, from
plowing and disking to the harvest, and the machines
used to accomplish these jobs.
   1. Agriculture — Juvenile literature.
2. Agricultural machinery — Juvenile literature.
[1. Agriculture. 2. Agricultural machinery]
I. Title.
S519.B87   1987   631   87-9333
ISBN 0-87614-314-1 (lib. bdg.)
ISBN 0-87614-493-8 (pbk.)

Manufactured in the United States of America

   3   4   5   6   7   8   9   10   97   96   95   94   93   92   91   90   89

To Dave and Virginia Cheeney, who have been
both an inspiration and an example to me

When you take a trip through the country, there is a lot to see if you know where to look. You might see a peaceful small town.

7

Or you might see a newborn calf being cleaned by its mother...

...or thousands of turkeys just looking sociable.

You will probably also see farmers at work in their fields. During the course of the year, farmers have a variety of jobs to do in order to grow, maintain, and **harvest,** or gather in, crops. Spring is the beginning of the growing season for many farmers. This is when the farmer **tills,** or prepares the fields, by **plowing** and **disking**.

Plowing turns over the topsoil and buries the weeds and stubble from last year's crops.

Disking breaks up the soil after plowing to prepare and smooth the field for planting the new crop.

The tractor is the most-used piece of farm equipment because it is used to pull so many other pieces of farm machinery. No matter what other equipment farmers have, they almost always have at least one tractor and sometimes even two or three.

Tractors come in all sizes, shapes, and colors. Until the 1950s, almost all planting, tilling, and harvesting equipment was attached to and powered by tractors. Even though many of today's farm machines run under their own power, the tractor is still a very important piece of farm machinery.

Tractors usually last a long time. Many tractors built in the 1940s and 1950s are still being used today.

Old farm machinery is very different from modern farm equipment.

This Case tractor, which was first built around 1900, was one of the biggest and most powerful tractors of the time. It burned wood to generate the steam that powered its engine. The wheels are made of iron, and each rear wheel weighs over 1,200 pounds. It probably cost about $2,500 when it was new.

This John Deere tractor is one of the largest tractors available today. It weighs 22,500 pounds and has a 230-horsepower engine. The tires can cost up to $500 each. The total price of this tractor is about $80,000.

This is an all-wheel drive tractor. That means that it pulls with all of its wheels instead of just the back ones. This tractor bends in the middle so the farmer can steer it. A farmer can use an all-wheel drive tractor with four, six, or eight tires. A tractor like this can plow as much in one day as an older, more conventional tractor is able to plow in four days. With an all-wheel drive tractor, today's farmer can plow in ten minutes what would have taken a farmer with one horse all day to do!

Here are two more conventional-looking tractors. The rear wheels of both of these tractors are the power wheels. The front wheels are smaller and are used to steer the tractor.

Some tractors can be used with up to three or four rear tires. This tractor is being used with three tires. Only one tire will fit in the **furrow**, or trench made by a plow, and the extra tire on the left gives added traction so the tractor won't slip.

After the fields are prepared, farmers use the tractor to pull a **seed drill** or **seeder** that plants the seeds. The seeder shown here plants crops that grow in rows such as corn, soybeans, and sunflowers. The center bin holds the seeds, and the side bins hold the **fertilizer**, a substance that is added to the soil to make it more productive. The poles at the sides of the seeder are row markers. The farmer uses them as guides to keep the rows straight. A different kind of seeder is used to plant grains like wheat and barley and grasses like alfalfa and clover that do not need to be planted in rows. This seeder **broadcasts**, or scatters, the seeds over the field.

With a little sunshine and a little rain, the plants soon start to pop up. Another growing season on the farm has begun.

When the crops start to grow, so do some unwanted plants called **weeds**. Farmers must get rid of weeds so that they do not crowd out crops by competing with them for sunlight, water, and nourishment from the soil.

One of the ways farmers control weeds is by **cultivating**. A cultivator digs up the weeds between crop rows.

Another way to control weeds is to spray them with **herbicide**, or weed killer. Each of the people riding on this machine has a hand sprayer to spray the weeds with a herbicide.

This odd-looking machine is a weed-spraying machine. It does not need to be hitched to a tractor because it has its own engine.

Spraying for weeds is usually done by machine. Hand spraying is done only when a farmer wants to put herbicide on the weeds and not on the crop.

Farmers often plant several different crops in the same field to reduce **soil erosion**. Soil erosion is when the topsoil becomes worn away by water and wind. To control this process, farmers usually plant strips of grass crops that hold the soil in place between strips of cultivated crops like corn that are planted in rows of loose soil. This is called **strip farming**.

After crops reach a certain height, they are too big to fit under a cultivator. Farmers can still control the weeds by spraying, but usually they don't do very much with the crops from this time until harvest, when the crops are ready to be picked. During this growing season, wonderful patterns are created by the crop rows in the rolling fields.

23

In order to have food for their livestock throughout the winter, many farmers make **hay** two or three times during the summer. Hay is usually a combination of cut and dried grasses, alfalfa, and clover.

A **mower** behind the tractor cuts the hay crop. After the alfalfa, clover, and grasses have been cut, they are referred to as hay.

The farmer then rakes the hay into rows with another attachment on the tractor called a **side-delivery rake**.

A **baler** makes the round bales. It can make up to 60 bales per day, which is about 45 tons of hay. Besides making bales, this baler also wraps and ties twine around each bale. Some balers make square bales and can pitch the bales into a wagon after they have been formed and tied.

Bales of hay are often left in the field to be picked up later. A **bale probe**, which is a giant needle that attaches to a tractor, is used to pick up these bales and move them from the field to a storage area.

Besides baling hay, farmers sometimes chop up hay with a **field chopper** and store it in a **silo**. Both the bales and the chopped hay are used as feed for cattle and other livestock.

After a long summer growing season, the crops are ready for harvesting in the fall. They are harvested by machines called **combines**. These machines are called combines because they combine the operations of cutting the plants and separating the grain **kernels**, or seeds, from the stalks.

This combine is cutting wheat. It removes the wheat kernels and spits the **straw**, or empty stalks, out the back. The kernels go into a bin in the combine. The bin is the large box behind the driver. When the bin is full, the farmer unloads the wheat into a truck or wagon using the chute you see sticking out. The straw is later collected and used for livestock bedding.

Combines were first manufactured in the 1930s. Prior to that time, threshing machines, like the one shown here, were used. While early combines were pulled by horses or tractors, modern combines are self-propelled, air-conditioned, and often have two-way radios that farmers use to talk to other people working on the farm. Today's combines can harvest three times as much grain in a day as the older combines.

In late September and early October, corn is ready to be harvested. The combine used for harvesting corn is the same one that is used for wheat, soybeans, sunflowers, and other grains. The head in front of the combine is changed for different crops. The head for corn and sunflowers has slots for picking row crops. The head for wheat, oats, barley, and other crops that are not planted in rows is straight in front with cutters to cut the plants and paddles to feed the plants into the combine.

Besides growing crops to feed their livestock, some farmers grow **cash crops**, or crops to sell. Most farmers store the corn they need for their livestock in storage bins or in corn cribs on their own farms, and they take the rest of the corn by wagon or truck to an **elevator** where grain is stored until it is shipped out to the final buyer.

When soybeans are ready to be harvested in the fall, the plants are very delicate. Even the wind can cause the beans to fall to the ground.

Harvesting soybeans is similar to harvesting wheat. The soybean plants are cut by the combine and pushed by paddles into the machine. Then the beans are separated from the stalks and collected in the bin.

Soybeans are probably used in more food products than any other farm crop with the possible exception of sugar beets. They are also used for livestock feed after the oil has been taken out.

In addition to harvesting soybeans, corn, and wheat, farmers also harvest vegetables in the fall. In this picture, carrots are being harvested. There are two different kinds of carrot pickers. The carrot picker above picks the carrots out of the ground by the greens, or tops.

The tops are cut off, the dirt is shaken loose, and the carrots go by conveyor over to the truck that is driven alongside the picker. When the truck is full, the carrots are taken to a buyer, who either resells them or processes them for use in food products.

This carrot picker cuts off the tops first. Then it scoops the carrots out of the ground, shakes off the dirt, and conveys the carrots to the wagon being pulled next to the picker. Farmers use the same kinds of pickers to harvest parsnips, turnips, and rutabagas as they use for harvesting carrots.

After the harvest, farmers spend a lot of time repairing and servicing their machines in preparation for the coming year. This is a very important job since throughout the planting, growing, and harvesting seasons, modern farmers rely on their machines to get the job done.

Glossary

bale: a compressed and bound package of material such as hay

baler: a machine that forms bales of hay

bale probe: a giant needle that is attached to a tractor and used to move hay bales

broadcast: to scatter seed in a field

cash crop: a crop that a farmer grows to sell

combine: a machine for harvesting grain that combines the operations of cutting the plants and separating the grain kernels from the stalks

cultivate: to dig up and loosen the soil around growing plants

disk: to prepare the soil for planting by cutting the soil with rotating metal disks

elevator: a tall warehouse used to store grain

fertilizer: a substance such as manure or a mixture of chemicals used to make soil more productive

field chopper: a piece of equipment used to chop up crops for livestock feed

furrow: a groove made in the ground by a plow

harvest: to gather in a crop

hay: grass or other green forage plants that have been cut and dried for livestock feed

herbicide: a chemical used to kill weeds

kernel: the seed of a grain plant

mower: a machine with a sharp, moving blade used to cut grass or other plants

plow: to work the soil by turning over the top layer

seed drill: *see* seeder

seeder: a piece of equipment used to plant or broadcast seeds

side-delivery rake: a piece of equipment that is attached to a tractor and used to rake cut hay into rows

silo: a tall, cylindrical structure where livestock feed is stored

soil erosion: the wearing away of topsoil

straw: the stalks left after the kernels have been harvested from grain plants

strip farming: a method of planting crops to reduce soil erosion in which different kinds of crops are grown across narrow strips of the same field

till: to prepare a field for planting a crop by turning, cutting, or leveling the soil

weed: any unwanted plant, especially one that crowds out more desirable plants

windrower: a self-propelled machine that can cut and rake a hay crop at the same time